PREPPING

The Survivalist Guide to Natural Calamities,

Wars, and Economic Turmoil

I0503412

Fhilcar Faunillan

TABLE OF CONTENTS

Introduction

I want to thank you for getting a copy of this book, *"Prepping: The Survivalist Guide to Natural Calamities, Wars, and Economic Turmoil."*

This book will provide you with useful information about the basics of "Prepping" so that you and your family will be ready no matter what fortuitous events may come your way.

Being too lax is not at all encouraged, nor is being too paranoid. Prepping is something that your family should practice, and your children are not an exception. Of course, it would greatly help if you know how to swim in times of flood. You should also have a sturdy house that can withstand harsh weather conditions. But besides that, you need to develop other prepping skills to be able to survive the aftermath.

Much as we do not want these unfortunate events to come our way, we must be ready, as nothing beats an ant who thinks of its future, especially when the rainy days come. Prepping should not only be a mindset but a lifelong practice because we will never know exactly when a calamity will strike.

.

Part 1

Prepping is Not Just for the Paranoid

CHAPTER 1:

Why Prepare?

Whether you prepared for it or called it nonsense, we're all happy that the world didn't end, as the Ancient Mayans said it would, in 2012. That being said, we are not off the hook as we could still face some disasters in the future.

Tornadoes and hurricanes could come into town and literally storm your life, or let's say Country A invades Country B for supposedly humanitarian purposes. Shhhh, don't tell. Armed soldiers are on your doorstep searching for something you don't have, and out of frustration, they just bomb your house. The same place whose mortgage you haven't even fully paid off yet. Next thing you know, you end up roaming the streets.

Maybe the economy could plummet on one dreadful night, and the next day, millions of people all over the world become jobless.

What then? Well, you need to prepare even before all those misfortunes could happen. You need to ensure that you have access to basic necessities in order to survive. This book will guide you through what you need to have on hand for different scenarios.

Let's start with why you need to prepare for the possibilities of natural disasters, wars, and economic turmoil. You may ask me, isn't that too much negative thinking? Can't we just stay as glass-half-full kind of people and assume that nothing as drastic as tsunamis wiping off an entire coastal town could occur?

No. Review your history and tell me that events like volcanic eruptions, devastating hurricanes, extreme global financial crises, and world wars don't happen. You can't. Catastrophic events are an unfortunate risk of human existence. Or perhaps, couldn't you just prepare after an announcement regarding, let's say, an incoming storm has been released? No. Just no. Waiting until the last hour to plan for a disaster isn't enough planning.

Suppose you're living in a country where almost nothing ever happens, good for you. But there are others out there whose countries experience twenty typhoons a year or fifty earthquake shocks annually. Even if natural catastrophes aren't the usual headlines in your local newspaper, the fact that it isn't unthinkable should give you a clue that it may very well happen to you too.

One day, your city may be in a state of emergency because the volcano lying dormant nearby has finally woken up. And in this case, you can find the word convenience in the dictionary but nowhere else in real life. The modern technology that you have come to depend on? The 24/7 store two blocks away that's usually your hunting ground whenever you're in the mood for some Toblerone? I'm sorry, but you can kiss them goodbye and miss them like hell.

Nowadays, we have come to depend greatly on what's instant and convenient. We have neglected the pursuit of knowledge and independence for the sake of comfort and immediate gratification. We twist a knob on the stove, and voila, a fire appears before our eyes. We pick up our phones, and pizza is on the doorstep half an hour later. Admit it. We can't perform simple daily activities without the help of electricity, instant food, and running water. Do you know how to start a fire with just matchsticks and wood? Do you know how to prepare food that wasn't sealed in a can or displayed in some fancy packaging?

You have to learn how to do these things because the time will come when these skills will be needed, and if you're not prepared beforehand, you'll be the first to suffer or, worse, to die.

The problem is, most people are misinformed and uneducated about emergency preparations. Not only do we remain clueless about the right things to do in case of unforeseen disasters, but we're also ambivalent about the topic of learning and educating ourselves. What we have now may not always be there.

Prepping is necessary because if there's one thing hurricanes like Katrina and Sandy have taught us, it's that nature is a bitch, and it doesn't play favorites. Poor or rich, you may end up lining for rations of food and water. It doesn't matter if you're the prettiest girl in town. You would have to endure not showering for a week because water will be scarce.

You don't get a say about your chances of being a victim of circumstances, but you can do something about your survival skills if disaster does strike.

CHAPTER 2:

What is Prepping?

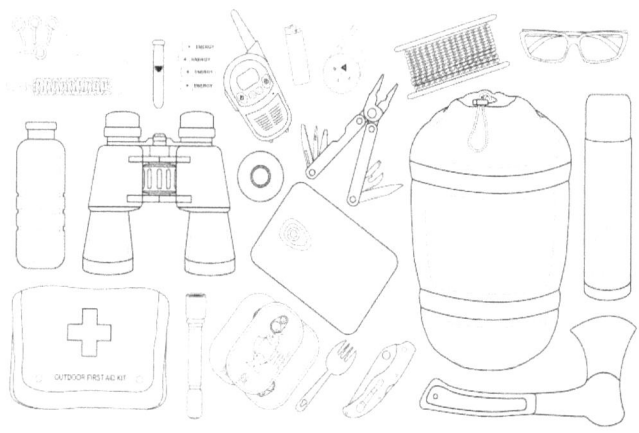

Prepping is not a fad, nor is it a novel undertaking. Generations before, it was a standard practice for every household to prepare for worst-case scenarios. Pantries were stocked with food, and cabinets were filled with supplies and other essential goods. It was only when people grew more dependent on the technology and services of other people that this practice became rare. Moreover, our increasing expectation that the government can provide us with whatever we demand comes to a faulty abandonment of self-reliance and independence.

Prepping is simple. Contrary to popular belief, it doesn't demand that you invest all your time, money, and effort in ensuring that you are bulletproof. It doesn't mean that you have to spend every single second of your life fretting over an impending disaster. Instead, it is about enjoying your life as it is today while you prepare for what might come tomorrow.

The common conception people have of preppers is that they're paranoid, tinfoil hat-wearing individuals who create bunkers filled with supplies. In fact, there are different levels. It can range from just a simple survival-kit preparation to full-out bunker creation, which is usually what laypeople had in mind when thinking of prepping. No matter what level of prepping you're willing to undertake, it is a given that you'll benefit from a couple of distinct advantages prepping can provide you in case of an emergency.

Prepping works on the idea that if ever an emergency is declared, people will rush out of their houses, literally like there was no tomorrow, towards the nearest grocery and hardware stores to stock up on basic necessities. But there is only so much stock a store can provide, and at the end of the day, shelves will be emptied, and others will have nothing to bring home because demand won't match supply in times like this. As a prepper, why are you better off?

For one, if you're already stocked beforehand with what you need to survive, you won't have to worry about rushing with other people anymore to get your hands first on basic necessities and risk getting yourself trampled.

Furthermore, in cases when there is no warning at all and the calamity strikes, you can rest assured that you have what you need even if stores and medical facilities aren't accessible to you. You don't have to waste your time and energy fighting with others over the last can of pork and beans, and instead, you can focus your energy on other important tasks that you have to do.

The bottom line is, prepping is a kind of mindset. It is about assuming responsibility for one's own welfare and independence from others. It also advocates self-sufficiency. People who shy away from preparing for disasters live in the false assumption that necessities will always be provided to them by the government. They live their lives in comfort, and they have a tendency to be lax because they are so dependent on the thought that help will surely come their way.

CHAPTER 3:

What Should You and Shouldn't Expect?

To survive during any kind of disaster whether natural or man-made, there are a few things that you need to be aware of. Not realizing some facts could be of a great disadvantage for you, so it will be better to keep in mind the following:

1. **Expect that public utilities will be closed down and will fail to work.**

Water, electricity, communications, among others, won't be yours for the taking. Usually, the supply or service to these basic needs would be cut off during this time, and announcements regarding when they will be turned back on won't be available any time soon.

2. **Expect that food will no longer be on store shelves.**

Once information about an incoming disaster has been disseminated, people will run like hell to the nearest store and buy everything they can. Even during calamities without warning, expect that a surge of people will be competing against each other over a bottle of water. Food will run out quickly, and if you're not prepared beforehand, you can find nothing on the display shelves.

3. **Expect that people will fight over the basic necessities.**

One of the benefits of preparing ahead of time would be escaping this debacle. Calamities are desperate times that would call for desperate measures. You can't expect order in an event when the survival of an individual and their family would be the topmost priority. Grocery stores could be the stage of the greatest fights ever, and there's nothing you can do about that. People would fight over what they need to survive.

4. Expect the worst.

In any kind of disaster, you have to think of the worst-case scenarios that could happen and prepare for them. You can always hope for the best, but prepping is better done when your mindset is hinged on all the necessities for you to survive could be difficult to access. This way, the moment disaster strikes, you are well-equipped and ready to respond accordingly.

5. Don't expect for the government to address your personal issues.

If you have trouble getting the government to hear your concerns on a typical day, what are the chances that they will attend to your needs quickly during a disaster especially when many people are affected? It is likely that they will need time to be able to send help for everyone. This is exactly why prepping heavily emphasizes on self-reliance and independence.

You can't expect any system to be there for you and answer all your problems. In cases such as massive calamities, there are millions of people that need attention. If you want to survive, you better do something about it yourself instead of waiting for other people to come and rescue you.

Part 2

Food Prepping for Everyone

CHAPTER 4:

Food Prepping

Sorry, man. Most likely, you won't be able to tell your mom to cook you her perfect pancakes and buy you Snickers in the middle of the night at the nearest 7/11 store. In times of disaster, you have to make do with what's available around you.

During natural catastrophes and emergencies, people usually react with fear and panic. They will go to stores and clean the shelves of every canned food that they could think of. This is a good move since even if the danger doesn't happen, people would still have a stock of food available for their families for the days to come. But as a prepper, you can have a more varied selection of food. It allows you to plan and carefully select the food you want because you are not rushing to buy things. Plus, when others are going crazy over at the cashier, you're not missing anything as you concentrate on last-minute concerns like making sure your family is well-informed and ready about whatever's coming.

The general idea of food prepping is stocking up on food that can last for a longer period and can sustain you for months, if necessary. Food is vital, as three weeks without them could be fatal. So, you have to make sure that your pantry has enough supply.

Buy food that doesn't need refrigeration, such as canned meat, vegetables, and tuna. It would also be good to have sugar, tea, rice, bread, oil, powdered milk, and pasta, among others. Look out for sales in the stores nearby and buy bulks of products at lesser prices.

You would still need cooking in some cases, which means you need water and fuel to cook your food, like rice and instant noodles.

Freeze-dried food and Meal Ready-to-Eat (also known as MRE) are also some of your best options. They come in sealed packages that are ready for immediate eating. Designed for the military, MREs are the ultimate instant food, as you would only need to heat them for a few minutes before they're ready for consumption.

When talking about shelf-life, freeze-dried food would have around ten to twenty years compared to MREs that only have three to five years of shelf life. Freeze-dried food is, therefore, better since they wouldn't require replacement. However, if you still want to consider buying MREs for your survival food storage, include them in your short or medium-term storage and replace them from time to time.

CHAPTER 5:

Food Stockpiling

Determining what to buy can be mind-boggling, given that there are thousands of food choices out there. The best course of action would be to stockpile food that is dense in nutrients. Avoid investing in chips and other food types that don't really hold nutritional value. There is no excuse for compromising your health, and cans of tuna and preserved vegetables would serve you so much better than 10 bags of chips.

The key is to focus on what you need instead of what you want. If you are clamoring for snacks or sweets, go for protein bars and chocolates. Chocolates, dark ones, in particular, can give you a quick source of energy and other benefits. Plus, they don't take up too much space.

The next step in stockpiling is actually building it. This part can unnerve some people, and they clamor to buy 100 cans of pork and beans all at once. Don't get ahead of yourself. You don't need to buy all 100 cans in one purchase. And 100 may be too much if you're only prepping for yourself. Take an accurate and honest assessment of your needs and base your choices on that evaluation. Plan out your meals. You can limit the amount of food you buy for a meal based on the calories or base it on serving sizes.

After planning and whenever you're grocery shopping, double the items that you usually buy, especially the non-perishable ones; let us say that you usually purchase 4 cans of pork and beans; make it 8. This way, you'll be building your stockpile quickly.

Once you have your stockpile sitting on your shelves, ensure that they do not pass the expiration date. If no disaster has hit, eat something from your supply to avoid waste. Rotate and replace your stock and always put the newest items you have bought in the back so that what you use are those you have bought first.

Another important note to keep in mind is to buy food that you can afford. Do not push your financial limits by buying 8 cans of pork and beans when your wallet can't stretch far enough. Also, if you're prepping for others (e.g., family members), you should take into consideration their preferences.

It is also recommended to stock a wide variety of food that offers both nutrition and satisfaction to avoid food fatigue. Avoid stocking up on only one type of food because sooner or later, you will not be able to stomach eating the same kind of food over and over again, and it would just go to waste. Another disadvantage of eating only one kind of food is that it may lack some nutrients that are needed by the body.

CHAPTER 6:

Cooking Food without Power

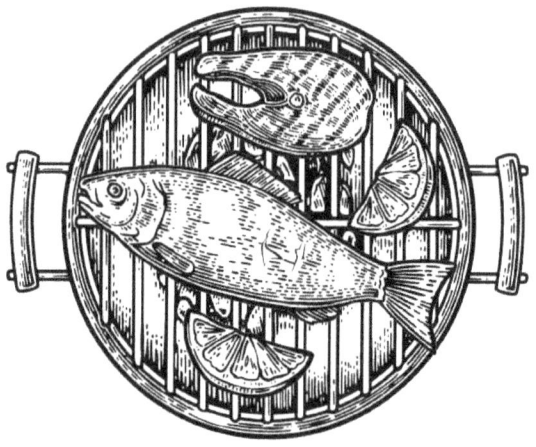

There will always be a higher chance of having a power outage during disasters, so don't depend too much on electricity when cooking. There are alternative ways to cook food; listed below are just some of them:

1. **Open fire**

   This method can be challenging, and it takes some practice to succeed in cooking with an open fire. However, this is the most common alternative heat source for cooking food and can produce excellent results when done properly. You have to have a lot of wood to keep an open fire going. Kettles, iron skillets, and Dutch ovens, among others, can be utilized to cook just about anything that you'd want to over an open flame.

2. Volcano stoves

Volcano stoves can be an ideal alternative to regular stovetop since they are very easily stored as they are collapsible. Furthermore, volcano stoves use three different types of fuel. Namely wood, charcoal, and propane, so you're not limited to just one fuel type. The efficiency of volcano stoves in terms of cooking food is something to be lauded at as well.

3. Grills

If you have the option of staying at home and aren't forced to live on the streets, you can still make use of your BBQ grill to cook your food. Just be sure that you also have at hand the necessary fuel (e.g., wood, charcoal, propane). The problem with grills is that some are not portable, so if you need to move away from your home, you can't bring it with you, and you'd have to learn another method of cooking food without power.

4. Camp stoves

Suppose you've got a camp stove lying around, then good! But you can't use them inside your shelter, and you would need to stock up on fuel (i.e., propane). Propane bottles usually last around just an hour, so you'd have problems with storage and expense.

5. Stove-in-a-can

A stove-in-a-can is a portable and lightweight gadget that you can include in your emergency kit. It is relatively inexpensive, and it runs on replaceable compressed fuel cells with a long burning life.

6. Rocket stoves

Learning how to cook on rocket stoves is relatively easy. This alternative cooking tool runs on a wood-burning, high-temperature combustion chamber where heat shoots out towards your food. An advantage a rocket stove has over the others is its extreme efficiency. However, you need to procure a supply of wood for fuel, and it's not that portable.

Part 3
Water Prepping

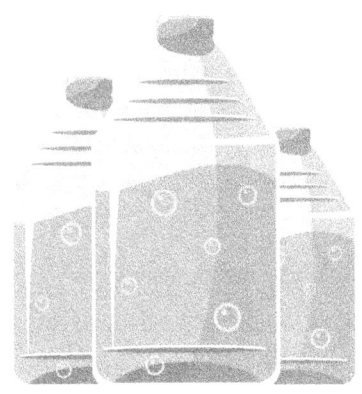

CHAPTER 7:

Sources of Water

Water is a highly underrated necessity. A lot of people usually prefer other kinds of beverages. But when disaster comes, water will be more appreciated than ever.

At present, you might think it is in excess, but in reality, especially in situations like a disaster, water can become so limited that people end up without any access to a source. Unless you live near a body of fresh water (e.g., river, lake), finding a sufficient amount that can last you for months can be a challenging feat.

Wars – especially major ones – are not easily resolved. After natural disasters, it can take a considerable amount of time before services in the communities can run again. Every day can be a bad day, and you will never know when things will be stable again.

At some point, you may run out of supplies, one of which you can't live without. Water is a vital necessity for human life. Therefore, the lack of it must not be dismissed as trivial. This section will discuss how you can find other water sources when the time comes the water you have stored has all been depleted.

1. **Rainwater**

   Make use of rainwater as it can be a source of potable water. You can use basins or buckets to catch it directly. You'll have some to use for bathing, washing your clothes, and drinking. Just remember to purify it first before drinking.

2. **Wells**

   Wells are excellent sources of water since they are sustainable. Know your place and find out if there is a well nearby that you can get water from.

3. Rivers and streams

Not all water coming from natural sources like streams and rivers is safe to drink. One sip from one wrong source, and you'll end up floating downstream. There are huge chances that your streams and rivers have been polluted with toxic waste. Serious ailments can develop from drinking unpurified water, and natural bodies of water are rich with bacteria, chemical run-off, parasites, and other waste products that could be fatal for you. There is no exception to the rule of purifying your water first before drinking it.

4. Salt and brackish water

Salt and brackish water are better left alone. You would need extreme purification before they can be used for drinking. However, you can always use them for other household purposes like cleaning and flushing the toilet.

CHAPTER 8:

Clean Water

Having water, more specifically, having safe water, is a basic need. In our preoccupation with providing ourselves with the necessities to survive, it will be easy to forget things such as food and water safety. Save yourself from experiencing unnecessary and cumbersome ailments while you are in the midst of a disaster and stock up on safe and purified water.

1. Boiling

If you don't have any water purification tools on hand, you can always go back to the basics and boil your water for at least 5 minutes to kill the harmful pathogens that would make your water unsafe for drinking.

2. Microfilters

Microfilters can be used to treat water as they are capable of removing bacteria from it. As bacteria and protozoans are the most common organisms that can be found in one's drinking water, it is a must to remove them. Microfilters with pore sizes of no larger than 0.3 microns are required for successful filtration. Glass fibers, plastics, and ceramics could be used as filtration media.

3. Chemical disinfectants

Chlorine, for one, can be used as an agent to treat water. Its effect is quick and it's readily available to anyone. But like other chemical disinfectants, chlorine doesn't hold any long-term effect. After some time, water will become contaminated again because of prolonged storage. This is not advisable for drinking purposes.

4. Conventional household bleach

Bleach that is used in the household can also be utilized to purify water. For one gallon of water, you can add 16 drops of regular household unscented liquid bleach to sanitize it. But, you can only use this for cleaning purposes, this is not advisable for drinking.

CHAPTER 9:

Water Storage

The water supply can be shut off without any announcement of when it will return. Especially during natural disasters, you would have to rely on your own water sources to avoid dehydration and perform basic household activities.

As for the amount of water that you should store, it is recommended to keep a gallon of water for every person per day for at least three days or how many days you are afraid the water supply won't be back on. There should also be enough reserved water for household purposes such as cleaning, bathing, and cooking. You must take into consideration your norms and standards. If you think a gallon of water per day for one person won't be enough because he uses a lot of water regularly, you should store more.

Do not be afraid to have too much water. The more you have, the better.

Now, your next concern should be on where to store your water. Bottles are the most obvious choice for water storage. However, buying in bulk can be expensive, so you have the option to store your water on an available container that you have. But it would be best if you remembered to take on the necessary precautions to maintain the cleanliness of your water and your container.

Here are some things to keep in mind when storing water:

1. **Use containers that are food-grade in storing water.**

   When you use other kinds of containers other than that of food-grade in handling water, you run the risk of dangerous substances and chemicals leaking into your water. In addition, do not put your water in metal containers since the metal will corrode. Opt for glass, plastic, or stainless steel instead.

2. **Do not use containers that have been previously used for another kind of beverage.**

Avoid using containers that have previously contained milk or juice, for example. With plastic bottles, in particular, substances from the previous beverage will stick to the plastic no matter how much you clean it. Your juice, for one, contains sugar that will trigger bacteria growth, so don't put your water into its container.

3. Remember to clean your containers.

Before storing your water in the containers, make sure that they have been sanitized well. Wash them with hot water and soap and rinse very well before filling them with water, and don't forget to keep your fingers away from the top of the bottle while you're filling it up.

4. Rotate and check your water regularly.

It is ideal to regularly check your stored water to ensure that it's still potable. Even though you've met every necessary measure to make it completely drinkable, there are a lot of ways in which your water can end up contaminated. If you need to clean the containers, wasting water would be a horrible idea. Instead of dumping it, you can use the water for different household chores like cleaning and laundry.

You can also purify the water again to make it safe for cooking and drinking.

Part 4 –
Shelter Prepping

CHAPTER 10:

Common Mistakes When Building a Shelter

Here are the common mistakes that people make when building their own shelter:

1. Too big and open

When it comes to space, people usually make their shelter big and open. There are no issues with your shelter's size as long as you don't make it open enough for a hyena or a mugger to come in. You have to make sure that your door is not wide open to protect yourself from an invasion of unwanted elements.

2. An accident in the making

Perhaps due to inexperience and misinformation, people make the mistake of building shelters that can collapse on their inhabitants. Before you try to create one, you need to educate yourself first on what is required to build a strong one. Make sure that you know how to set up the structural foundation because it ensures stability.

3. Overestimating warmth

You were dead tired as you were building your shelter, so you got sloppy. That is no excuse. Or perhaps you were injured, and you could not do any better. That is acceptable, but the fact still remains that you'll probably turn to a human Popsicle.

If you thought that putting one flimsy blanket on the floor to keep you warm is enough to shield you from the cold of the night, then you're – literally – dead wrong. In building your own shelter, one of the utmost priorities is insulation. Don't just think that because your shelter looks good, it is going to feel fine. Before you go to sleep, ensure that your shelter is blizzard and hurricane-proof. In case it becomes too hot for you, you can open it a little bit to let in some air.

4. Wrong choice of elevation

A mistake that people often commit is building their shelters on really high grounds. Higher elevations will not keep you warmer as they will expose you to the wind. Moreover, fire is more difficult to build and maintain because the heat will be carried away from you by the wind, and fuel will be burned faster. Drier areas protected from the elements on lower grounds will always be your best option.

How to Build a Basic Shelter?

The shelter is one of the basic needs as it provides necessary elements for a person's survival. It provides protection to a person and also an avenue to keep one warm. In unfortunate scenarios where you have to move from your home or have lost one, you can create or buy various types of basic shelters in your local store.

1. Tents

Tents are not meant to be permanent shelters but can be used as temporary methods of keeping away from harmful elements. To an extent, it provides some security against animals and will keep you warm. Preppers usually choose tents when anticipating short-term emergencies and include commercially produced tents in their survival kits.

It would be best if you and your family know beforehand to put up a tent. You can spend a weekend away from home, like camping, to learn how to live in a tent without any access to modern technology and luxuries. This kind of activity will help develop your survival skills and lets you have proper mindset.

2. Caves

If you know your terrain beforehand, you can look for caves where you can stay temporarily. Remember to check the place for any insects or wild animals that can pose a danger to you before you inhabit any place.

3. Using materials in the wild

Especially in cases when you end up in the wild, you can make use of what nature can provide. Tree limbs, leaves, and rocks can be utilized to become structures of a make-shift shelter.

Part 5

Other Important Things

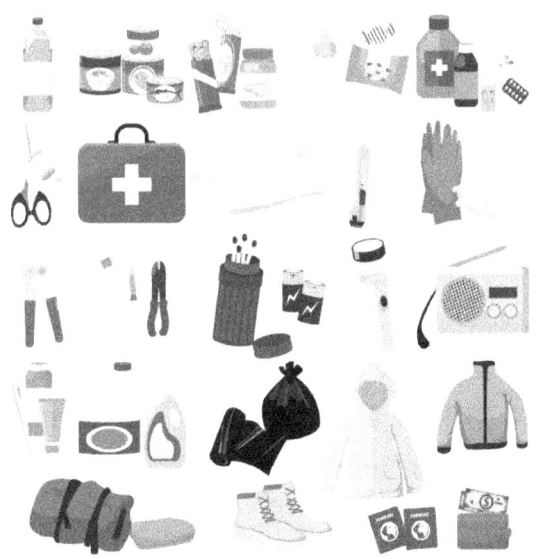

CHAPTER 11:

Building Your Own Fire

It is most likely that you will lose power, especially in post-disaster situations. Electrical lines will be down, and your electric stove is pretty much useless in this situation. Moreover, let us not neglect the fact that you might end up in the wild or without a home, so appliances would not be within your reach even if they are working.

One of the most fundamental skills that you have to learn in preparation for disasters is fire-building. Fire is a man's best friend. Unless you can stomach sushi and raw meat for every meal of every day and somehow have the refrigeration for that, you need fire to cook and boil your water. Most importantly, you need it to keep you warm and alive.

Moreover, it would also provide illumination for you and a way to ward off animals.

Since you can never tell what type of disaster may strike, you need to educate yourself on the different methods of building your own fire to have a reservoir of skills to exhibit when the situation arises. And mind you, another good thing about preparing for this task is that you do not have to stress yourself out by rubbing two stones together to create a spark. Now, you can equip yourself with the necessary tools to help you build a fire easily.

1. Matches

Using matches is the old-age method of starting a fire, so they should be part of your survival kit. Avail of waterproof matches if you can find any or put the regular ones in waterproof containers to ensure that you can have a working fire in whatever weather you may be experiencing. It would also be better to choose wooden matches over paper ones.

2. Firesticks

There are a lot of firesticks sold at camping stores. Firesticks are relatively easy to carry since they are

small, and there are types that you use even if your
kindling is damp.

3. Lighters

It would also be handy to keep a couple of lighters
with you. They are very easy to use and can be reused,
unlike matches. Butane ones are ideal for windy and
wet situations, so that they may be preferable. You
have to make sure that you bring extra refills for your
lighters.

4. Ferrocerium

Ferrocerium produces intensely hot sparks when
scraped against other metals and other textured
surfaces. Ferrocerium rods are awesome additions to
your fire-starting kit before they can work in wet or
dry environments. They also come in portable sizes.

CHAPTER 12:

Maintaining Proper Hygiene

It would be so easy to forget sanitation and proper hygiene in a time when your every thought is zeroed in on surviving. But it is an issue that you should not neglect because diseases and illnesses could develop from improper hygiene.

I hope that by now, you already know that contact with fecal matter (poo) – can be detrimental to your health. Dysentery, diarrhea, cholera, etc., are just some of the possible diseases you can acquire without proper sanitation.

Since water would be an issue and you don't want to waste it on flushing the toilet, human waste would have to be let loose somewhere else.

1. **Composting toilets**

   Composting toilets are available at online or local camping stores. They are ideal for the proper and easy disposal of wasters. Some models separate liquids from solids, and your waste ends up in a sealed bucket and will be treated with bleach to get rid of bacteria and other disease-causing parasites. This way, you can confidently dump your waste into your latrine without any fear of contamination—one thing that you must remember - **never dispose of your waste near a water source**.

2. **Buckets**

   Disaster time is no fancy time. So, if you do not have any other option or if composting toilets are too expensive for your budget, you can always opt for a bucket lined with a garbage bag.

   You would need a large bucket, perhaps a five-gallon one, thick garbage bags, and lime. Line your bucket with the heavy-duty garbage bag, and after doing your business, cover it with chlorinated lime and place the lid of the bucket back on. Once the garbage bag starts to get filled, seal it up, take it out of the bucket, and replace it with a new one.

3. Latrines

Dig a latrine nearby to manage your waste. But again, remember not to do this near a potential water source. If there is a body of water nearby, dig your latrine at least 200 feet away from it. For a latrine that multiple people can utilize, dig one that is at least one and a half feet wide and 1 foot deep. Stock up on chlorinated lime so you can cover your waste with it every time you defecate. Other alternatives to lime would be wood ash or kitty litter. Additionally, think twice about digging a latrine if you live in a place prone to flooding, as making one would not be a wise decision.

4. Cat holes

Cat holes sound cute, right? They also look cute and are one of the easiest and fastest ways to maintain hygiene quickly. Dig a small hole of about 8 inches deep and away from a water supply and do your business. Do not forget to cover the hole up once you are done.

Aside from proper ways of waste disposal, there are more things you need to keep in mind regarding the issue of hygiene.

Add to your stockpile list a lot of toilet paper. Secure a roll of toilet paper for one person per week and adjust it depending on the necessity. Males and females may differ in the amount they need, so consider when you are stocking up. In addition, you have to procure garbage bags for each person for their individual sanitation needs.

Chlorinated lime or other alternatives is also a must in your kit and for females, remember to stock up on feminine hygiene products like sanitary napkins and tampons.

If possible, use biodegradable bags as much as you can and ensure that they won't easily break down while lined up in your bucket.

CHAPTER 13:

Learn First Aid

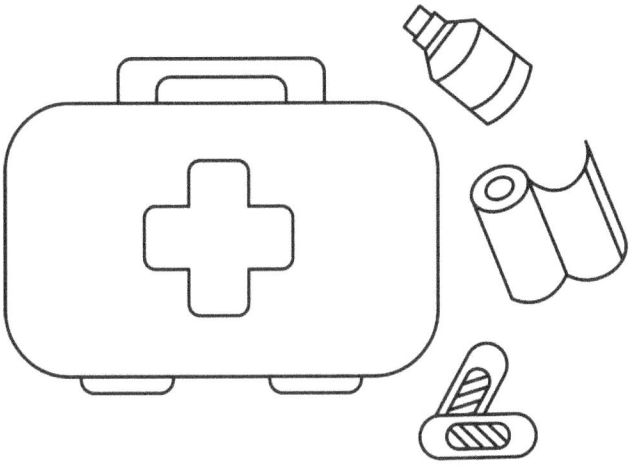

First aid is very important. You just can't get wounded and expect to have it healed after licking it or waiting for the natural healing process to work. In situations like these, you must be on top of your game and waiting for a wound – especially a debilitating one – to heal without treating it would be highly disadvantageous to you.

Aside from storing up on the necessary components of your first aid kit, you also need to either have a good manual indicating the uses and application of each item in your kit or, preferably, learn about them firsthand. Listed below are the things that must be present in your first-aid kit to help you beat the odds during calamities. There are pre-stocked first-aid kits available from stores, so you can purchase them and add your medications and other additional supplies for wound management into the kit.

1. Latex or sterile gloves
2. Antibiotic ointment
3. Adhesive bandages in different sizes
4. Sterile dressings for bleedings
5. Thermometer
6. Soap and other cleansing agents
7. Face Towels
8. Burn ointment
9. Decontaminants
10. Prescription medications of members you are prepping for (e.g., asthma inhalers, insulin, anti-allergy)
11. Anti-diarrhea medication
12. Pain relievers
13. Laxatives
14. Antacids
15. Lubricant
16. Scissors
17. Tweezers

If you happen to be out there in the wild, do not forget that medicinal plants can help treat a wound. But, you should familiarize yourself with their specific uses. For example, blackberries can do a lot for your stomach aside from just satiating it. Blackberry leaves are helpful in treating diarrhea. And if you want to repel insects and bug bites, procure some lavender or neem leaves. Always practice the necessary care when you're using plants from the wild. If you are not a hundred percent sure about the purpose of a particular plant, don't risk it. The same goes for eating these plants.

One of the myths about outdoor survival that can be quite fatal is the claim that you can eat and use everything in the wild. With proper knowledge, you will know that there are poisonous plants and mushrooms out there that will ease your suffering, only because they will send you to the afterlife, which is sort of what we're trying to avoid here.

CHAPTER 14:

Preparing a Bug-Out Bag

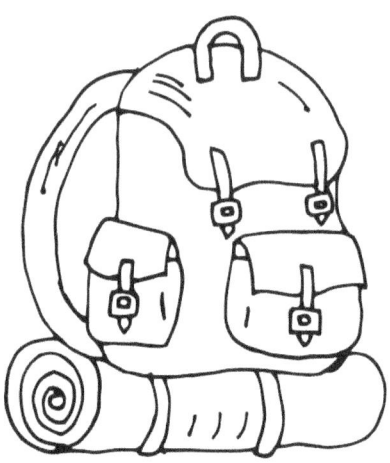

In the absence of the luxuries we may be used to in the modern world, you have to make do with what is available in times of disaster.

Every prepper should have a Bug-Out Bag prepared. A Bug-Out Bag or BOB is one that contains your kits and various items that will help you survive in the face of natural calamities, wars, and other life-threatening disasters. It can be personalized to be suitable for a specific period.

BOB's are designed to be portable so that you can take them with you in the event that you have to leave your house in an instant. There is no fixed formula for the perfect BOB as it would depend on your location and the potential threats you could be facing.

A lot of situations could demand that you leave your home. From natural catastrophes to social uprisings, these events can come unexpectedly. So, it would help if you were on your toes and be prepared for anything. Besides being well-informed and experienced in different survival skills, you also need to possess a collection of supplies and tools to aid your survival. And keeping these tools in one bag is an efficient way to get the hell out of an emergency in one piece. You won't need to run around the house looking for what you have because everything is already in one place.

I have listed below the things you need to consider in preparing your BOB.

1. Purpose

In preparing your BOB, you need to have a clear purpose in mind. What is your BOB for? Will you use it for a long time, or is it just moving from one place to another? This part is an important step in getting your BOB ready because it will tell you how much or how little of an item you need to bring.

2. Travel details

The distance and the terrain you'll be traveling in are vital points to take note of. When you try to move away from people from different countries shooting the hell out of each other or, let us say, running away from an incoming tsunami (God bless you), speed is significant. Your BOB shouldn't be so heavy that you can't take a few steps forward. The heavier it is, the slower you can run away from trouble. You also have to add into the equation your physical capabilities. As a person who hardly ever does any form of physical activity, a forty-pound of BOB will probably be the death of you.

If you're traveling long distances, you can aim for BOBs weighing a total of twelve pounds. If you want to cross between points A and B, a twenty-pound BOB will do.

This is where trial runs would be most beneficial for. Take out your BOBs for trial runs and see if you are comfortable with its weight. Be honest regarding what your body can take, and don't push it.

3. Shelter and warmth

Always make sure that your BOB contains your tools for building a shelter. Shelter preparation is highly essential because ending up frozen to death due to inadequate protection is exactly what you're trying not to do. Fact is, you can survive three weeks without food and three days without water, but a night with hypothermia will kill you instantly. Hone your skills in building a shelter, purchase stable and sturdy items, and don't go cheap on this part because while you are building a faulty tent, a bear is probably on its way to eat you.

4. Hydration

Carrying bottles filled with water is a good idea, but it can be exhausting to bring a pack heavy with water. So, what you need to secure in your BOB are your water purification tools to filter any water you may find in any bodies of water nearby. However, you need to find out first if there are sources of water where you are heading because otherwise, your tools will end up useless; in that case, you would have to carry bottles of water to remain hydrated.

5. Food

The question is this: how long do you see yourself out there? Answer that question and make it a base for how much food you need to bring with you. To save space and weight, people usually include energy bars and canned food in their BOBs. But this doesn't mean that they are the only things that you'll be eating. They are intended to give you energy and sustenance at the beginning, enough that you can engage in other methods of acquiring food such as hunting, fishing, foraging, and trapping. But take note that these methods can only be done with the appropriate prior knowledge and skills. Your tools, I reiterate, will be useless if you do not know how to use them.

Additionally, when putting MREs and freeze-dried food in your BOB, take note of the amount of space and weight that you can afford to lose to your canned and heavy goods.

6. Hygiene and First Aid

Have an ample supply of the medicine that you need. Prepare what you need for your allergies because finding a drug store will be a challenge when you're out there trying to survive. Make sure as well that your first aid kit is well-equipped. Don't just settle for a kit filled with Band-Aids because that can hardly help when you're bleeding or poisoned or stabbed to death. Even the smallest of cuts can turn really complicated because of infections, so you need to invest in proper treatment.

Stock up your BOBs with your hygiene kit. This consists of shampoo, soap, toothbrush, washcloth, menstrual hygiene products, deodorant, toilet papers, etc.

7. **Fire Starting**

If you've been driven into the outdoors because of wars or natural disasters, you are as good as dead without fire and warmth for a significant period. It would be best if you learned to start a fire under any weather condition. Prepare your matches, tinder, fire sticks, lighters, or anything else that will help build a fire.

8. **Light**

Preparing lights, especially portable ones, is a must because it is most likely that power will be cut off during disasters, and it would be hard to recognize a thing in front of you at night when there's no illumination. A solar-charged light that doesn't need any battery would be ideal, as well as candles. You can also secure camp lights or headlamps. Invest in quality lights to not have any regrets when night comes.

9. Clothing

Clothing is not optional and is definitely an item that you need to add to your BOB. You need to prepare extra sets of clothes and ensure that they're appropriate for the weather. For colder climates, jackets and long pants are necessary. And in no circumstance must you wear shorts no matter the weather. Even when it's hot out there, your legs need protection from the heat of the sun. Not to mention that wearing long pants will decrease your skin's chances of getting irritated with allergens and getting bitten by insects.

In addition, keep your feet from injuries by wearing a good set of shoes. Change your socks as often as possible and wash them properly. If there's enough space in your pack, prepare two sets of shoes that you can alternate depending on the situation and location. Choose footwear that you're comfortable walking in. There's no time to think about what's fashionable as you get away from toppling trees or dangerous country invaders.

10. Bug-Out Bags

Don't be too frugal when it comes to your BOB, and never settle for a cheap backpack with terrible quality. Trust me, they will tear apart, and you'll end up with your things on the ground.

You need to invest in a good backpack because it is where your tools for survival are contained. Resort to heavier packs because they tend to be thicker and more durable compared to ultra-light ones. And remember: take out your BOBS for trial runs in order to assess its effectiveness. These trial runs will inform you of potential problems with the items you have prepared.

Part 6

Financial Prepping

CHAPTER 15:

Is Your Money Ready?

Much has been said about preparing for doomsday caused by wars or natural disasters. Books have been written about ways to find sources of water and stockpiling food in order to survive the whole duration of the calamities. However, what people fail to acknowledge is another kind of event that would disrupt the food supply and trigger social unrest – an **economic crisis**.

It's not out of the question that an economic instability large enough to cause everyday living dysfunctions could happen. Financial markets are highly volatile and sensitive, while stock markets are easily affected by politics and social events. The rising price of oil, which can be attributed to political unrest, would lead to corresponding increases in prices of daily goods and impact people's ability to live comfortably at their means.

In reality, a lot of personal and small-scale financial catastrophes happen every day. Institutions close because of bankruptcy, homes foreclosed, and people lose their jobs because companies are in huge debt. In cases like these, people are left without an income to sustain the needs of their families. Imagine what would happen if a global financial crisis occurs.

Financial prepping is just as vital as making it a point that you have enough supplies to last you during the rainy days because, like natural calamities, economic turmoil could also rob you of the chance to survive.

Therefore, it is important that you also devise a strategy, so you won't get blind-sided in case of major economic disasters.

1. Simplify finances

Stop spending on unnecessary and extravagant things. When you already have a functional one, a new vacuum cleaner will not do you any good when you have nothing to eat. As much as possible, learn to live a minimalist but functional life. Think several times before you purchase any product and ask yourself if you really need that item. Ultimately, the aim is to reduce your expenditure and increase your savings so you can afford to acquire basic necessities during difficult times.

2. Eliminate any kind of debt, especially the adjustable ones

Debts will bite you in the ass most when you don't want it to. Credit cards, for example, exist in an environment that depends on inflation. When a financial crisis hits, banks could be forced to raise fees to make up their losses, and you'd have to shoulder that increase.

3. Invest in what can make you self-sufficient

Spend money only on necessities. As food can become scarce or inaccessible during whatever type of crisis, it would be a great move to invest in them as well as in water and fuel. You can opt to allot some money on growing plants and fruit trees to have a steady food supply to put on your table. Additionally, you can purchase essential items that will give you purchasing power if it ever comes to a point when the economy devolves into a bartering system. Case in point, storing a bag of sugar can perhaps guarantee you a month's worth of rent.

4. Save

Saving is one of the most important steps in prepping for a financial disaster. When saving, you essentially choose between severe frugality when the demand arises and gradual/moderate frugality. You can be forced to significantly limit your budget and adjust when a crisis hits because you'd be obliged to. Now, this would be extremely hard because you're exposing yourself to a practice that you haven't tried before or you don't have a background in. Imagine that from $4 per meal, you are pushed to reduce it to $2 because of financial constraints. Not only would this scenario be hard to stomach at first, but it would also take a toll on your well-being. The better option would be to embrace moderate frugality little by little. As early as today, learn and try to live within your means. Try to assess what you can live without and build yourself a monetary cushion by spending only what is necessary and comfortable enough.

Don't shock yourself into doing drastic measures like doing without resources that are previously available to you. Let's say that you have been used to meals that cost a hundred dollars. Don't immediately jump into reducing it to twenty-five in one blink. Make yourself comfortable with a lesser and lesser budget for food in every meal until you find the right balance between satisfaction and frugality. Start building an emergency fund.

Saving doesn't also mean that you are required to give up everything that you have today. Learn to know what is essential and prioritize practicality over extravagance.

1. Budget

Budget and keep track of your income and expenses. Those that are non-essential to daily living should be cut out of your budget. Proper allocation to different essential items such as food, utilities, transportation, savings, among others, must be observed.

2. Sell non-essential items

For additional income, you can also get rid of items that you no longer use. Make an inventory of things that are redundant or would only add additional cost when utilized. You can put them up during a garage sale or sell them on eBay.

3. Learn DIY skills

DIY skills will not only come in handy for survival, but they can also cut you a lot of costs if you learn to do things that you would initially pay for. Learning them would mean that you stop paying exorbitant fees to other people for executing tasks that you can very well master on your own. Some of the requisite skills that a prepper should know include first aid, car and bicycle maintenance, appliance repair, food preservation, water treatment, self-defense, hunting and fishing, home maintenance, outdoor survival, weapons training, etc.

Now, the next question is where to learn these skills. There are various options open for you. You can always enroll in your community college or avail yourself of community resources. There are a lot of communities that offer free courses on emergency preparedness.

The Internet is another good option as it is a hub of knowledge. Videos and detailed manuals on various topics are available online for you to study, and the number of skills you can hone through the Internet is limitless.

If you're a more traditional kind of learner, you can always go to the library and borrow books on the topics you're interested in.

Conclusion

It's about time that we bring back prepping as a mainstream mindset. Unlike other trendy ideas out there, this can do us a lot of good because it increases our chances of surviving in the event of natural disasters, wars, and financial breakdowns.

We must change our lax attitude and our beliefs that nothing terrible can happen to us because it can and maybe it will. Circumstances can be very random and out of our control, but at least, by preparing beforehand, we say whether or not we will end up as victims.

Keep in mind your primary needs: water, air, food, and shelter. Make it a point that you have a plan for all these, and you know what to do to have a sustainable supply of these necessities when the demand arises.

Just remember that by prepping, it's not only you that can benefit but also your loved ones. After reading this book and taking this survivalist guide to heart, don't forget to share your newly acquired knowledge with family members and friends. Education is the first step to prepping, and you will be doing them a great favor by informing them of the things they should do to be ready.

I don't have any qualms about being a prepper. People laugh at extreme preppers and find them ridiculous, wondering over the use of their bunkers and the storing of several canned goods in their pantries. But if you think about it, these people that are laughed at are more likely to survive any disaster because they are well prepared.

So, I guess the next question is when you should start prepping? How about right at this moment? There is no excuse for delaying such an important task. Make it a part of your lifestyle - live it today so you can live tomorrow.

www.ingramcontent.com/pod-product-compliance
Lightning Source LLC
Chambersburg PA
CBHW070915180526
45168CB00005B/2025